PERENNIAL MATH

CONTEST PROBLEMS

2018-2019

Dr. Sylvia Dean was a gifted specialist and administrator for 26 years in Huntsville, Alabama. She created Perennial Math in 2011 upon completion of her doctoral degree. She has hosted many robotics, technology, and mathematics competitions throughout the years and has tirelessly worked on providing educational opportunities to students around the world. It has always been her ambition to leave students with the feeling that "math is cool."

These problems are challenging and students should know up front that scoring a 3 out of 6 on any of these tests is extraordinary and above average. Great problem-solvers learn persistence and appreciation for challenge. Good luck as you encourage your students to take the quest to become great mathletes.

Rookie
level

Grades 3-4

Name________________________________

PERENNIAL MATH

TEST #1

November 2018 Rookie Level

1) 1, 2, 4, 7, 11, 16, ___, ___, ___ What is the sum of the three missing numbers in this sequence?

2) Cameron has 8 fewer marbles than Roger. Jeff has 6 fewer marbles than Roger. Roger has more than 10 marbles. Altogether the boys have 37 marbles. How many marbles does Cameron have?

3) If an equal number of 2's, 4's, and 6's are added together to get 96, what is the total number of 6's used in the sum?

4) Using the information from the diagram, solve for the value of one turkey minus one hat.

5) A number such as 121 is a palindrome because it reads the same forwards and backwards. How many palindromes are between 100 and 450?

6) Donna has quarters, dimes, nickels, and pennies in her jar. She had ten of each coin in her jar. How many ways could she make 63 cents?

PERENNIAL MATH

TEST #1

November 2018 Rookie Level Answers

1) 1, 2, 4, 7, 11, 16, ___, ___, ___ What is the sum of the three missing numbers in this sequence?	88
2) Cameron has 8 fewer marbles than Roger. Jeff has 6 fewer marbles than Roger. Roger has more than 10 marbles. Altogether the boys have 37 marbles. How many marbles does Cameron have?	9
3) If an equal number of 2's, 4's, and 6's are added together to get 96, what is the total number of 6's used in the sum?	8
4) Using the information from the diagram, solve for the value of one turkey minus one hat. 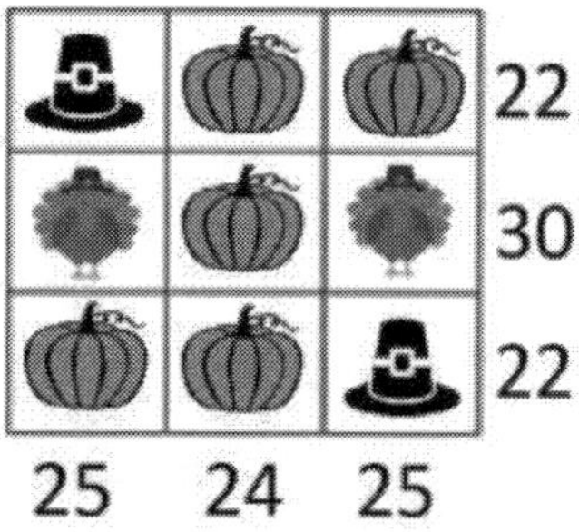	5
5) A number such as 121 is a palindrome because it reads the same forwards and backwards. How many palindromes are between 100 and 450?	35
6) Donna has quarters, dimes, nickels, and pennies in her jar. She had ten of each coin in her jar. How many ways could she make 63 cents?	22

PERENNIAL MATH

TEST # 1

November 2018 Rookie Level Solutions

1) 1, 2, 4, 7, 11, 16, ___, ___, ___ What is the sum of the three missing numbers in this sequence? This sequence has the numbers adding +1, +2, +3, +4 and so on. The next three numbers are: (16 + 6 = 22), (22 + 7 = 29), (29 + 8 = 37).

1, 2, 4, 7, 11, 16, 22, 29, 37

So the sum of the three missing numbers is 22 + 29 + 37 = 88

88

2) Cameron has 8 fewer marbles than Roger. Jeff has 6 fewer marbles than Roger. Roger has more than 10 marbles. Altogether the boys have 37 marbles. To solve for how many marbles Cameron has, make a chart. Start with Roger having 11 marbles and increase that number until you have a total of 37 marbles. Cameron has 9 marbles.

Roger	Cameron	Jeff	Total (need 37)
11	3	5	19
12	4	6	22
13	5	7	25
14	6	8	28
15	7	9	31
16	8	10	34
17	9	11	37

9

3) An equal number of 2's, 4's, and 6's are added together to get 96. To find the total number of 6's used in the sum, make a chart. Start with 1 of each number and increase that amount until the sum of them is 96. There are eight 6's in the sum.

2's	4's	6's	2 x 2's	4 x 4's	6 x 6's	Sum(96)
1	1	1	2	4	6	12
2	2	2	4	8	12	24
3	3	3	6	12	18	36
4	4	4	8	16	24	48
5	5	5	10	20	30	60
6	6	6	12	24	36	72
7	7	7	14	28	42	84
8	8	8	16	32	48	96

Note: Alternatively, 2 + 4 + 6 = 12 or 1 set and 96 ÷ 12 = 8 sets.

8

PERENNIAL MATH

TEST # 1

November 2018 Rookie Level Solutions

4) Use algebra to find the value of one turkey minus one hat.
Let P = pumpkin, H = hat, T = turkey. In the middle column, 3P = 24, therefore P = 24/3 or 8. In the bottom row, 2P + H = 22. Since P = 8, then 2P = 16. 22 – 16 = 6, H = 6. In the third column, H + T + P = 25. 6 + T + 8 = 25. T = 25 – 14 = 11. Now to solve for T – H, just put in the values and subtract. 11 – 6 = 5.

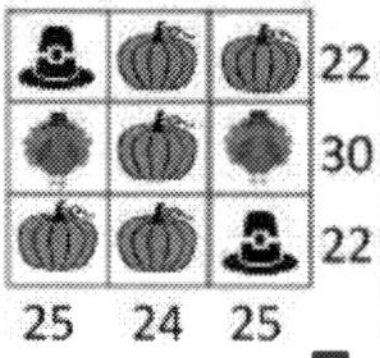

5

5) A number such as 121 is a palindrome because it reads the same forwards and backwards. How many palindromes are between 100 and 450?
The palindromes between 100 and 450 are: 101, 111, 121, 131, 141, 151, 161, 171, 181, 191, 202, 212, 222, 232, 242, 252, 262, 272, 282, 292, 303, 313, 323, 333, 343, 353, 363, 373, 383, 393, 404, 414, 424, 434, and 444.

35

6) Donna has 10 of each coin (quarters, dimes, nickels, and pennies) in her jar. Make an organized list to see how many ways she could make 63 cents. There are 22 ways to make 63 cents using 10 or less of each coin.

Q (.25)	D (.10)	N (.05)	P (.01)	Total (.63)
2	1	0	3	0.63
2	0	2	3	0.63
2	0	1	8	0.63
1	3	1	3	0.63
1	3	0	8	0.63
1	2	3	3	0.63
1	2	2	8	0.63
1	1	5	3	0.63
1	1	4	8	0.63
1	0	7	3	0.63
1	0	6	8	0.63
0	6	0	3	0.63
0	5	2	3	0.63
0	5	1	8	0.63
0	4	4	3	0.63
0	4	3	8	0.63
0	3	6	3	0.63
0	3	5	8	0.63
0	2	8	3	0.63
0	2	7	8	0.63
0	1	10	3	0.63
0	1	9	8	0.63

22

Name________________________________

PERENNIAL MATH

TEST #2

December 2018 Rookie Level

1) 372 - 273 + 89 =	
2) A miner sold 198 grams of gold on the first day, and 24 grams on the second day. On the third day he sold 18 grams less than he sold on the first day. On the fourth day he sold 16 grams more than he sold on the third day. How many total grams of gold did he sell altogether?	
3) Sally spray-painted the blocks below while they were sitting on the table. How many of the blocks only had 4 sides painted? 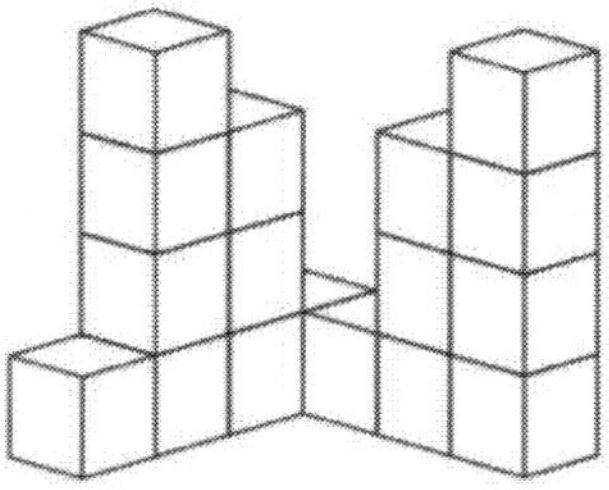	
4) Jack has 600 magic beans and he plans to sell 15 beans every day. If he starts selling his beans on a Monday what day of the week will he finish selling his beans?	
5) Max and Jack have 14 tickets together. Bob and Max have 20 tickets together. Jack and Bob have 12 tickets together. How many tickets do all three of them have altogether?	
6) Steve makes $3.75 per cut. If Steve is cutting the log into 9 pieces, how much money will he make?	

PERENNIAL MATH

TEST #2

December 2018 Rookie Level Answers

1) 372 - 273 + 89 =	**188**
2) A miner sold 198 grams of gold on the first day, and 24 grams on the second day. On the third day he sold 18 grams less than he sold on the first day. On the fourth day he sold 16 grams more than he sold on the third day. How many total grams of gold did he sell altogether?	**598**
3) Sally spray-painted the blocks below while they were sitting on the table. How many of the blocks only had 4 sides painted? 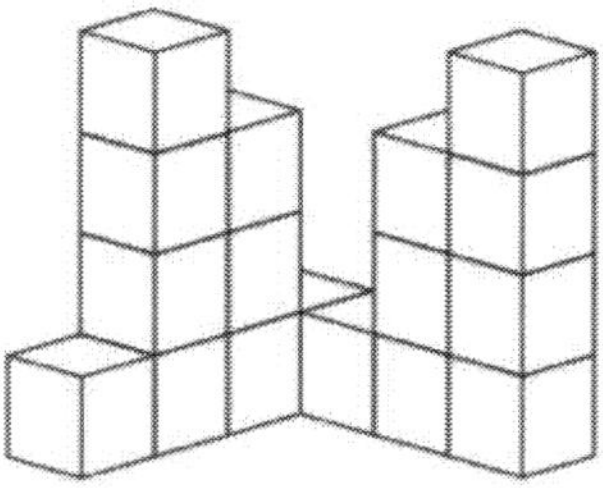	**3**
4) Jack has 600 magic beans and he plans to sell 15 beans every day. If he starts selling his beans on a Monday what day of the week will he finish selling his beans?	**Friday**
5) Max and Jack have 14 tickets together. Bob and Max have 20 tickets together. Jack and Bob have 12 tickets together. How many tickets do all three of them have altogether?	**23**
6) Steve makes $3.75 per cut. If Steve is cutting the log into 9 pieces, how much money will he make?	**30**

PERENNIAL MATH

TEST #2

December 2018 Rookie Level Solutions

1) 372 – 273 + 89 =

Solve using two steps

372 – 273 = 99

99 + 89 = 188

188

2) A miner sold 198 grams of gold on the first day, and 24 grams on the second day. On the third day he sold 18 grams less than he sold on the first day. On the fourth day he sold 16 grams more than he sold on the third day. How many total grams of gold did he sell altogether? To solve add each day's sales.

198 + 24 + (198 -18) + (198 - 18 + 16) = 598 grams of gold.

598

3) Sally spray-painted the blocks below while they were sitting on the table. How many of the blocks only had 4 sides painted? Since Sally spray-painted the blocks while they were sitting on the table, the bottoms would not be painted. Blocks on the ends that do not have a block on top of them and only one block adjacent to them would have only four sides painted. The one block touching the table with only one block adjacent would also only have four sides painted. There are 3 blocks that had only 4 sides painted.

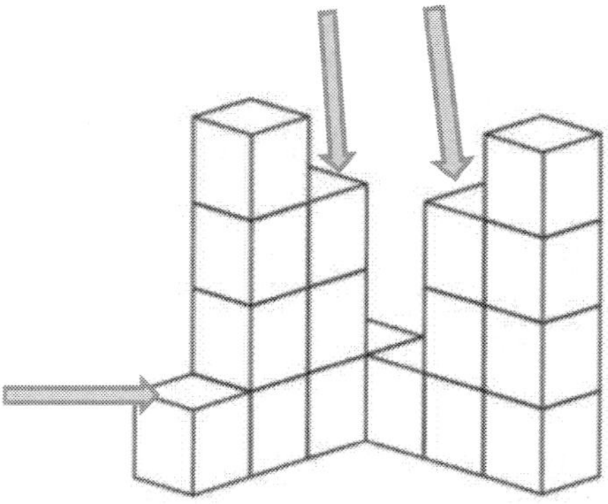

3

PERENNIAL MATH TEST #2

December 2018 Rookie Level Solutions

4) To find what day of the week Jack will finish selling his beans, first figure out how many days it will take him to sell 600 magic beans. If he plans to sell 15 beans every day, it will take him 600/15 = 40 days. Now divide 40 by 7 to see how many weeks it will take him. 40/7 = 5 weeks with a remainder of 5 days. Since he starts selling his beans on a Monday, five full weeks plus five days will end on a Friday.

Friday

5) Max and Jack have 14 tickets together. Bob and Max have 20 tickets together. Jack and Bob have 12 tickets together. Use algebra to find how many tickets all three boys have altogether. Let M = Max; J = Jack; and B = Bob.

$$
\begin{array}{rrrr}
 & M & + J & = 14 \\
B + & M & & = 20 \\
+\ B + & & J & = 12 \\
\hline
2B + & 2M + & 2J & = 46
\end{array}
$$

$$\frac{2B + 2M + 2J}{2} = \frac{46}{2} \quad \text{\{Now divide both sides by 2\}}$$

$$B + M + J = 23$$

23

6) Steve makes \$3.75 per cut. Steve is cutting the log into 9 pieces. To find how much money will he make, first figure out the number of cuts he will have to make to cut the log into 9 pieces and then multiply by \$3.75. To get 9 pieces, he will need to only make eight cuts. So, the total amount of money he make is \$3.75 x 8= \$30

1 cut = 2 pieces
2 cuts = 3 pieces
...
...
8 cuts = 9 pieces.

30

Name________________________________

PERENNIAL MATH

TEST #3

January 2019 Rookie Level

1) 9 x 8 = 27 + ___	
2) Calvin bought 89 candy bars at the Dollar Store. The candy bars cost \$1 each, but you could get 6 for \$5. What was the least amount he could pay for the candy bars without tax?	
3) Mitchell decided to lose weight. He wanted to lose 5 pounds a week until he got down to 160 pounds. Brooks started losing weight 2 weeks after Mitchell and he lost 3 pounds a week. If Brooks started at 175 pounds and got down to 160 pounds the same time as Mitchell, what was Mitchell's beginning weight?	
4) In this cryptarithm, each letter stands for a different digit. What is the value of SMORE? MORE MORE + SMORE 45571	
5) Mrs. Smith bought some number stickers at the store to put on each student desk. Each stickers only had one of the digits 0 through 9 on it. She numbered the desks consecutively starting at 1. Each digit costs 8 cents each. She paid a total of \$2.80 for the stickers. How many desks did she number?	
6) Brian is mailing a card to his grandmother. He only has 42¢ and 33¢ stamps and it costs 1.87¢ to mail the card. He wants to put the lowest amount of postage on the card. He had to waste ___cents of postage over the required amount.	

PERENNIAL MATH

TEST # 3

January 2019 Rookie Level Answers

1) 9 x 8 = 27 + ___	45
2) Calvin bought 89 candy bars at the Dollar Store. The candy bars cost $1 each, but you could get 6 for $5. What was the least amount he could pay for the candy bars without tax?	75
3) Mitchell decided to lose weight. He wanted to lose 5 pounds a week until he got down to 160 pounds. Brooks started losing weight 2 weeks after Mitchell and he lost 3 pounds a week. If Brooks started at 175 pounds and got down to 160 pounds the same time as Mitchell, what was Mitchell's beginning weight?	195
4) In this cryptarithm, each letter stands for a different digit. What is the value of SMORE? MORE MORE + SMORE 45571	41857
5) Mrs. Smith bought some number stickers at the store to put on each student desk. Each stickers only had one of the digits 0 through 9 on it. She numbered the desks consecutively starting at 1. Each digit costs 8 cents each. She paid a total of $2.80 for the stickers. How many desks did she number?	22
6) Brian is mailing a card to his grandmother. He only has 42¢ and 33¢ stamps and it costs 1.87¢ to mail the card. He wants to put the lowest amount of postage on the card. He had to waste ___cents of postage over the required amount.	5

PERENNIAL MATH

TEST # 3

January 2019 Rookie Level Solutions

1) 9 x 8 = 27 + ___ (multiply 9 x 8)
72 = 27 + ___ (subtract 27 from both sides)
72 - 27 = 45
The missing number is 45.

45

2) Calvin bought 89 candy bars at the Dollar Store. The candy bars cost $1 each, but you could get 6 for $5. To find the least amount he could pay for the candy bars without tax divide the 89 candy bars by 6. 89/6 = 14 with a remainder of 5. He paid 14 x $5 = $70 for 84 candy bars. Then he paid an additional $5 for the five remaining candy bars. $70 + $5 = $75.

75

3) Mitchell decided to lose weight. He wanted to lose 5 pounds a week until he got down to 160 pounds. Brooks started losing weight 2 weeks after Mitchell and he lost 3 pounds a week. Brooks started at 175 pounds and got down to 160 pounds the same time as Mitchell. To find what weight Mitchell started at, first solve for how long it took Brooks to lose his weight and then figure out how many weeks Mitchell had to lose his weight. If Brooks started at 175 and got down to 160, then he lost 175 - 160 = 15 pounds. If Brooks was losing weight at 3 pounds per week, then it took him 15/3 = 5 weeks to lose his weight. Now we know that Mitchell started losing weight 2 weeks before Brooks. So Mitchell lost his weight over 5 weeks + 2 weeks = 7 weeks time. Since Mitchell was losing 5 pounds per week, he lost 5 pounds x 7 weeks = 35 pounds. 35 pounds + 160 pounds = 195 pounds. Mitchell weighed 195 pounds at the beginning.

195

PERENNIAL MATH

TEST # 3

January 2019 Rookie Level Solutions

4) In this cryptarithm, each letter stands for a different digit. What is the value of SMORE?

```
  MORE
  MORE
+ SMORE
 45571
```

First solve for the 3 E's. Since their sum ends in 1, E must be 7 and a 2 carries over to the tens digit. 3R's plus 2 ends in 7, so R must be 5 and a 1 carries over to the hundreds digit. 3 O's plus 1 ends in 5, so O must be 8 and a 2 carries over to the thousands digit. 3 M's plus 2 ends in 5, so M must be 1. Since there's no carry, then S = 4. Therefore SMORE = 41857.

41857

5) Mrs. Smith bought some number stickers at the store to put on each student desk. Each stickers only had one of the digits 0 through 9 on it. She numbered the desks consecutively starting at 1. Each digit costs 8 cents each. She paid a total of $2.80 for the stickers. To find out how many desks she numbered first solve for how many digits she bought. $2.80 divided by .08 equals 35 digits. Now start numbering the digits that she would put on the desk. Desks 1-9 would be 9 digits. Desks 10-19 would be 20 digits. That would be 29 digits. So six more digits would be the numbers 20, 21 and 22. There were a total of 22 desks.

22

6) Brian is mailing a card to his grandmother. He only has 42¢ and 33¢ stamps and it costs 1.87¢ to mail the card. He wants to put the lowest amount of postage on the card. Make a table to find how many cents postage he had to waste over the required amount. He had to waste five cents.

0.42	0.33	Total Spent	Money over - 1.87
5	0	2.1	0.23
4	1	2.01	0.14
3	2	1.92	0.05
2	4	2.16	0.29
1	5	2.07	0.2
0	6	1.98	0.11

5

Name________________________________

PERENNIAL MATH

TEST #4

February 2019 Rookie Level

1) Sam has 10 quarters and 5 dimes. Sally has 5 quarters and 10 dimes. How many more cents does Sam have compared to Sally?

2) Solve these two problems. How much needs to be added to the smaller answer to make it 100 more than the larger answer?

$$\begin{array}{r} 753 \\ -\ 185 \\ \hline \end{array} \qquad \begin{array}{r} 478 \\ -\ 289 \\ \hline \end{array}$$

3) John starts a savings account with $100. Every week he adds money to his account. The first week he put in $8; the next week he puts in $3; and then he adds $5 the next week. If he continues rotating this amount $8, $3, and $5, then how many weeks will pass until he has $220 in his

4) In this diagram, what is the value of 1 star plus 1 pentagon?

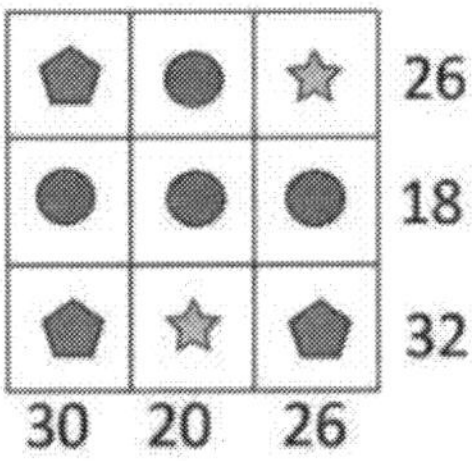

5) Peyton had 36 candy bars and he gave 24 away to his friends. He then ate 1/6 of the remaining candy bars. What fraction of his original candy bars does he have left? (Answer in simplified fraction form.)

6) Connie is putting her cookies in piles of 2 and there is one left over. If she puts them in piles of 3 there are two cookies left over. If she puts them in piles of 5 there are four cookies left over. How many cookies does Connie have if she has more than 35 but less than 65 cookies?

PERENNIAL MATH TEST #4

February 2019 Rookie Level Answers

1) Sam has 10 quarters and 5 dimes. Sally has 5 quarters and 10 dimes. How many more cents does Sam have compared to Sally?	75
2) Solve these two problems. How much needs to be added to the smaller answer to make it 100 more than the larger answer? 753 − 185 478 − 289	479
3) John starts a savings account with $100. Every week he adds money to his account. The first week he put in $8; the next week he puts in $3; and then he adds $5 the next week. If he continues rotating this amount $8, $3, and $5, then how many weeks will pass until he has $220 in his	22
4) In this diagram, what is the value of 1 star plus 1 pentagon? 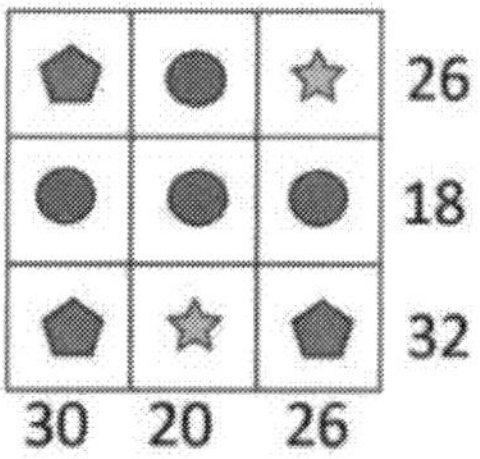 	20
5) Peyton had 36 candy bars and he gave 24 away to his friends. He then ate 1/6 of the remaining candy bars. What fraction of his original candy bars does he have left? (Answer in simplified fraction form.)	5/18
6) Connie is putting her cookies in piles of 2 and there is one left over. If she puts them in piles of 3 there are two cookies left over. If she puts them in piles of 5 there are four cookies left over. How many cookies does Connie have if she has more than 35 but less than 65 cookies?	59

PERENNIAL MATH

TEST #4

February 2019 Rookie Level Solutions

1) Sam has 10 quarters and 5 dimes. Sally has 5 quarters and 10 dimes. To find out how many more cents Sam has compared to Sally, first find out how much each has and subtract. Sam has 10(.25) + 5(.10) = 2.50 + .50 = 3.00. Sally has 5(.25) + 10(.10) = 1.25 + 1.00 = 2.25. Sam has more money so subtract Sam -Sally. 3.00 - 2.25 = .75. Sam has 75 cents more.

75

2) Solve these two problems. How much needs to be added to the smaller answer to make it 100 more than the larger answer?

$$\begin{array}{r} 753 \\ -\ 185 \\ \hline 568 \end{array} \qquad \begin{array}{r} 478 \\ -\ 289 \\ \hline 189 \end{array}$$

Since 189 is the smaller answer you will need to add to it to make it 100 more than the larger answer. The larger answer plus 100 is 568 + 100 = 668. Subtract the smaller answer from 668. 668 -189 = 479. You will need to add 479 to the smaller answer to make it 100 more than the larger answer.

479

3) John starts a savings account with $100. Every week he adds money to his account. The first week he put in $8; the next week he puts in $3; and then he adds $5 the next week. If he continues rotating this amount $8, $3, and $5, then how many weeks will pass until he has $220 in his account? To find out how many weeks will pass until he has $220 in his account, first subtract the original $100 deposit and from $220. $220 -$100 = $120. Then find out how many weeks it takes him by dividing $120 by $16 because every 3 weeks he has put in $8 + $3 + $5 = $16. $120/$16 = 7 with $8 remaining. Seven times 3 weeks is 21 weeks, plus an additional week to put in the remaining $8. It takes him 22 weeks total to save up to $220.

22

PERENNIAL MATH

TEST #4

February 2019 Rookie Level Solutions

4) 4) Use algebra to find the value of 1 star + 1 pentagon. Let P = pentagon; C = circle; and S = star. First solve for C. The middle row has 3C = 18, or 18/3 = 6, so C = 6. Next solve for P. The first column has 2P + C = 30. So, substitute C = 6 into this equation to get 2P + 6 = 30. 2P = 30 6 = 24. 2P = 24 , so P = 12. So, the value of 1 star + 1 pentagon is 8 + 12 = 20.

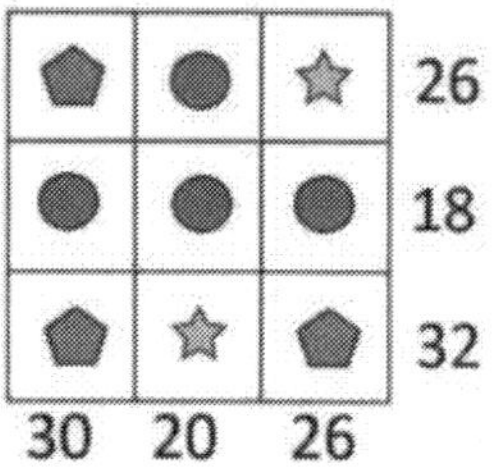

20

5) Peyton had 36 candy bars and he gave 24 away to his friends. He then ate 1/6 of the remaining candy bars. What fraction of his original candy bars does he have left? Subtract 36 24 candy bars he gave to his friends, so he had 12 candy bars left. He then ate 1/6 of these which is 12 x 1/6 = 2 candy bars. Then he would have 12 2 = 10 candy bars left. Convert this to a fraction and simplify. 10 out of 36 = 10/36. The answer must be in simplified fraction form, so 10/36 can be reduced to 5/18.

5/18

6) Connie is putting her cookies in piles of 2 and there is one left over. If she puts them in piles of 3 there are two cookies left over. If she puts them in piles of 5 there are four cookies left over. She has more than 35 but less than 65 cookies. Since she has one left over when she does piles of 2, it is an odd number. Make a list of the odd numbers between 35 and 65 and do elimination.

37, 39, 41, 43, 45, 47, 49, 51, 53, 55, 57, 59, 61, 63

Eliminate all the numbers that are multiples of 3, and then all the numbers that end in 5, because she has remainders when she puts her cookies in piles of 3 and piles of 5.

37, 41, 43, 47, 49, 53, 57, 59, 61

Now look for numbers that leave a remainder of 4 when divided by 5.

49, 59

Then out of these two numbers only 59 has a remainder of 2 when it is divided by 3. She had 59 cookies.

59

Name________________________________

PERENNIAL MATH

January 2019 Rookie Level

TEST #1

1) If you multiply 7 and 6 the answer is _____ more than if you add 7 and 6.

2) Caleb's mom is 38 and his dad is 44. After how many years will the sum of their ages equal 100?

3) If $a ⌂ b = (a \times b) + (a - b)$, then how much larger is 9⌂7 than 7⌂6?

4) The perimeter of a rectangle is 40 cm. If its width is 6 cm, what is the area of the rectangle?

5) Helen drove 480 miles to the beach. If she drove 60 miles per hour for half the trip and 80 miles per hour the other half, how many hours did it take her to drive to the beach?

6) Eric strung racquets from 10:53 AM to 5:05 PM today. At what time was he halfway done stringing racquets? __________PM

PERENNIAL MATH

TEST #1

January 2019 Rookie Level Answers

1) If you multiply 7 and 6 the answer is _____ more than if you add 7 and 6.	**29**
2) Caleb's mom is 38 and his dad is 44. After how many years will the sum of their ages equal 100?	**9**
3) If *a* ⌂ *b* = (*a* x *b*) + (*a* – *b*), then how much larger is 9⌂7 than 7⌂6?	**22**
4) The perimeter of a rectangle is 40 cm. If its width is 6 cm, what is the area of the rectangle?  6cm	**84**
5) Helen drove 480 miles to the beach. If she drove 60 miles per hour for half the trip and 80 miles per hour the other half, how many hours did it take her to drive to the beach?	**7**
6) Eric strung racquets from 10:53 AM to 5:05 PM today. At what time was he halfway done stringing racquets? __________PM	**1:59**

PERENNIAL MATH

TEST # 1

January 2019 Rookie Level Solutions

1) If you multiply 7 and 6 the answer is _____ more than if you add 7 and 6.
7 x 6 = 42. 7 + 6 = 13. So, 42 – 13 = 29.

29

2) Caleb's mom is 38 and his dad is 44. To find out how many years will pass until the sum of their ages equal 100, solve for their current age and subtract. 38 + 44 = 82. 100 – 82 = 18. Now divide 18 by 2 people, 18/2 = 9. In 9 years their ages will add up to 100. Check your work: 38 + 9 = 47. 44 + 9 = 53. 47 + 53 = 100.

9

3) If $a \bigcirc b = (a \times b) + (a - b)$, then how much larger is 9⌂7 than 7⌂6?
Solve both equations then subtract to find the difference.

9⌂7 = (9 x 7) + (9 - 7) =
63 + 2 = 65
7⌂6 = (7 x 6) + (7 - 6) =
42 + 1 = 43
So, 65 – 43 = 22

22

PERENNIAL MATH

TEST # 1

January 2019 Rookie Level Solutions

4) The perimeter of a rectangle is 40cm. If its width is 6cm, what is the area of the rectangle? First solve for the length of the rectangle by subtracting the two widths. 40 – 2(6) = 40 – 12 = 28. Divide by 2 since there are two length sides. 28/2 = 14. Now you can find the area of the rectangle by multiply length times width. 14 x 6 = 84cm^2.

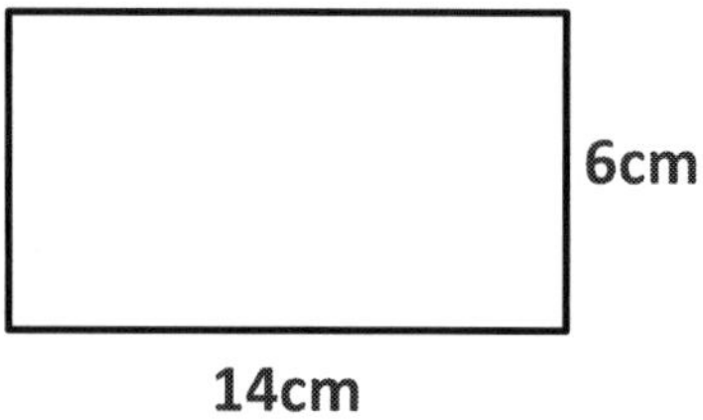

84

5) Helen drove 480 miles to the beach. If she drove 60 miles per hour for half the trip and 80 miles per hour the other half. To find how many hours it took her to drive to the beach, solve for both halves of the trip. The first half of the trip would be half the distance which is 480 miles/2 = 240 miles. If she drove 60 miles per hour, then 240 miles/60 mph = 4 hours. The second half or 240 miles she drove 80 miles per hour. 240 miles/80 mph = 3 hours. So 4 hours plus 3 hours is 7 hours. It took her 7 hours to drive to the beach.

7

6) Eric strung racquets from 10:53 AM to 5:05 PM today. To find what time he was halfway done stringing racquets, first solve for how many total minutes he worked and then divide in half. 10:53 to 4:53 is 6 hours. Then there are 12 minutes from 4:53 to 5:05. 6 hours x 60 minutes = 360 minutes plus the additional 12 minutes. So, he worked a total of 372 minutes. 372 minutes divided in half is 186 minutes. 186 minutes divided by 60 minutes is 3 hours with 6 additional minutes. Add this to his start time. 10:53 + 3 hours is 1:53. An additional 6 minutes would be 1:53 + :06 = 1:59.

1:59

Name________________________________

PERENNIAL MATH

February 2019 Rookie Level

TEST #2

1) 1000 – 109 – 19 – 9 =

2) If 18 doughnuts cost $7.20, how much would 24 doughnuts cost?

3) Jessica bought a square sheet of 144 stickers. The stickers were arranged in 18 rows of 8 stickers. How many stickers were not on the outside border of the page?

4) Cameron spray painted the blocks below while they were sitting on the table. What *fraction* of the blocks only had 2 sides painted?

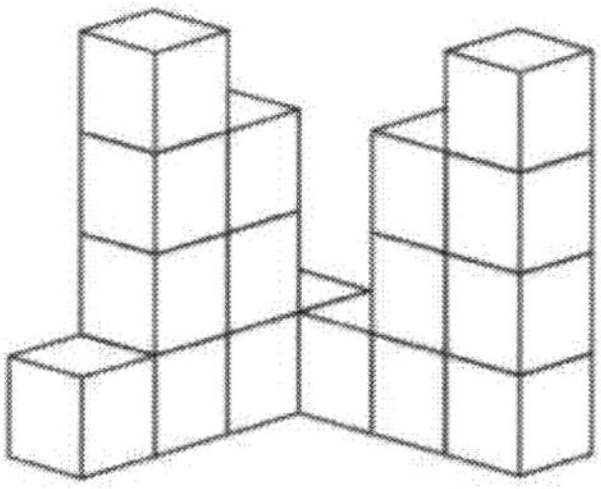

5) Melody and Debbie had 48 candy hearts together. Melody gave Debbie six of hers and then they both had an equal amount. How many candy hearts did Debbie have to begin with?

6) How many four-digit numbers can be made from the digits 1, 2, 3, and 4 if the digits can only be used once in the number?

PERENNIAL MATH

TEST #2

February 2019 Rookie Level Answers

1) 1000 – 109 – 19 – 9 =	863
2) If 18 doughnuts cost $7.20, how much would 24 doughnuts cost?	9.60
3) Jessica bought a square sheet of 144 stickers. The stickers were arranged in 18 rows of 8 stickers. How many stickers were not on the outside border of the page?	96
4) Cameron spray painted the blocks below while they were sitting on the table. What *fraction* of the blocks only had 2 sides painted? 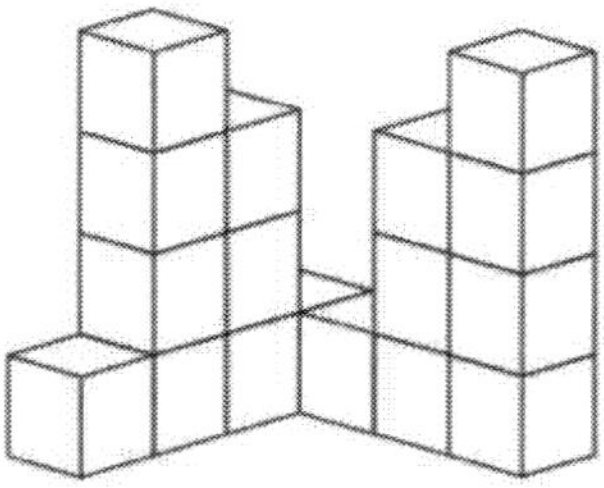	3/17
5) Melody and Debbie had 48 candy hearts together. Melody gave Debbie six of hers and then they both had an equal amount. How many candy hearts did Debbie have to begin with?	18
6) How many four-digit numbers can be made from the digits 1, 2, 3, and 4 if the digits can only be used once in the number?	24

PERENNIAL MATH

TEST # 2

February 2019 Rookie Level Solutions

1) 1000 – 109 – 19 – 9 =
1000 – (137) = 863

863

2) If 18 doughnuts cost $7.20, at the same rate how much would 24 doughnuts cost? Divide 7.20 by 18 to find out how much one doughnut cost. 7.20/18 = .40. Now multiply .40 by 24 to find out the cost of 24 doughnuts. .40 x 24 = 9.60

9.60

3) Jessica bought a square sheet of 144 stickers. The stickers were arranged in 18 rows of 8 stickers. How many stickers were not on the outside border of the page? You can make a drawing of the stickers and count them all or notice that if you subtract the top, bottom, and sides then the remaining stickers are in a 6 by 16 configuration. 6 x 16 = 96 stickers.

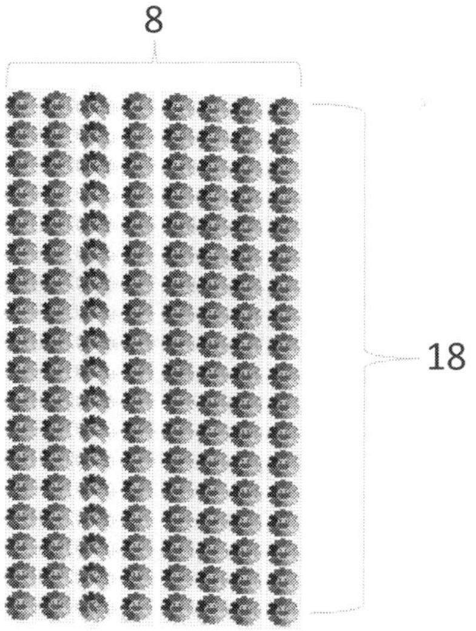

96

PERENNIAL MATH TEST #2

February 2019 Rookie Level Solutions

4) Cameron spray painted the blocks below while they were sitting on the table. What fraction of the blocks only had 2 sides painted? So the bottom blocks did not have paint on the side touching the table. If you count the total number of blocks, there are 17. There are 3 blocks that only have 2 sides painted because they have a block on top of them and blocks adjacent, plus the bottoms are touching the table. Therefore are 3/17 blocks with paint on only 2 sides.

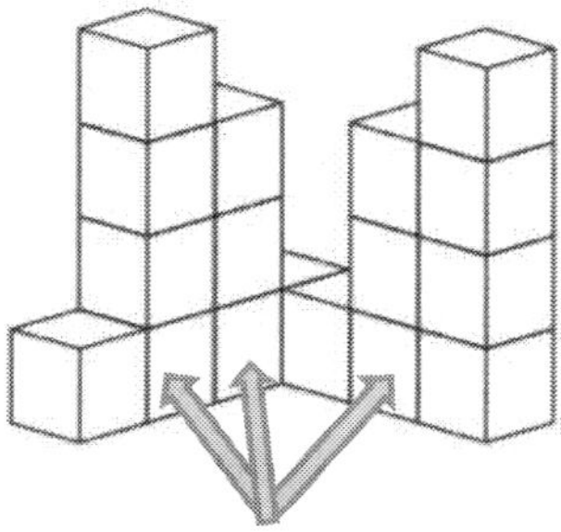

3/17

5) Melody and Debbie had 48 candy hearts together. Melody gave Debbie six of hers and then they both had an equal amount. To find out how many candy hearts Debbie began with, start by finding out how many both girls will have at the end and work backwards. They both will have equal amounts so 48/2 = 24 each. Since Debbie got six from Melody, subtract 6 from her 24. 24 – 6 = 18. Debbie had 18 hearts to begin with.

18

6) How many four-digit numbers can be made from the digits 1, 2, 3, and 4 if the digits can only be used once in the number? You can make an organized list and see that there are 24 four-digit numbers that can be made. Or there are 6 numbers that can be made that start with each digit. This will be the same pattern for each column of numbers when you start making numbers using a different starting digit. You will still only be able to make 6 numbers with each starting digit.

1234	2134	3124	4123
1243	2143	3142	4132
1324	2314	3214	4213
1342	2341	3241	4231
1423	2413	3412	4312
1432	2431	3421	4321

24

Name________________________________

PERENNIAL MATH

TEST #3

March 2019 Rookie Level

1) 20 + ___ = ___ x 5. What number goes in the blanks to make this a true statement?	
2) Jonathan had four quarters, sixteen dimes, and twenty nickels. He put his coins in four stacks so that all the stacks had the same amount of money. What is the least number of coins that he could put in any stack?	
3) Mike has 8 less cards than Bob. Jeff has 6 less cards than Bob. Bob has more than 10 cards. Altogether the boys have 43 cards. How many cards does Bob have?	
4) In this cryptarithm, each letter stands for a different digit. What is the value of CRATE? RATE RATE + CRATE 37428	
5) Tom was bored in class so he started writing out the counting numbers beginning at 1 and then 2 and so on. What was the 185th digit that he wrote?	
6) Gigi has a bag of 140 shells. The bag contains only limpets, cockles, and conchs. There are three times as many cockles as limpets. There are six times as many conchs as limpets. How many conchs are in the bag?	

PERENNIAL MATH

TEST #3

March 2019 Rookie Level Answers

1) 20 + ___ = ___ x 5. What number goes in the blanks to make this a true statement?	5
2) Jonathan had four quarters, sixteen dimes, and twenty nickels. He put his coins in four stacks so that all the stacks had the same amount of money. What is the least number of coins that he could put in any stack?	5
3) Mike has 8 less cards than Bob. Jeff has 6 less cards than Bob. Bob has more than 10 cards. Altogether the boys have 43 cards. How many cards does Bob have?	19
4) In this cryptarithm, each letter stands for a different digit. What is the value of CRATE? RATE RATE + CRATE 37428	32476
5) Tom was bored in class so he started writing out the counting numbers beginning at 1 and then 2 and so on. What was the 185th digit that he wrote?	7
6) Gigi has a bag of 140 shells. The bag contains only limpets, cockles, and conchs. There are three times as many cockles as limpets. There are six times as many conchs as limpets. How many conchs are in the bag?	84

PERENNIAL MATH

TEST #3

March 2019 Rookie Level Solutions

1) 20 + ___ = ___ x 5. What number goes in the blanks to make this a true statement? Use guess and check to find the answer. The number 5 makes this a true statement because 20 + 5 = 25 and 5 x 5 = 25. So 20 + 5 = 5 x 5.

5

2) Jonathan had four quarters, sixteen dimes, and twenty nickels. He put his coins in four stacks so that all the stacks had the same amount of money. To find the least number of coins that he could put in any stack, first solve for how much money he has and then divide by four to see how much money has to be in each stack. 4 quarters is 1.00. Sixteen dimes is 1.60. Twenty nickels is 1.00. This is a total of 1.00 + 1.60 + 1.00 = 3.60. 3.60 divided by 4 is .90 in each stack. Now the least amount of coins possible in a stack would be 3 quarters + 1 dime + 1 nickel or 5 coins.

5

3) Mike has 8 less cards than Bob. Jeff has 6 less cards than Bob. Bob has more than 10 cards. Altogether the boys have 43 cards. To find how many cards Bob has make a chart and start with Bob having 11 cards. Increase Bob's cards until the total sum is 43. Bob has 19 cards.

Bob (B)	Mike (B - 8)	Jeff (B - 6)	Total (43)
11	3	5	19
12	4	6	22
13	5	7	25
14	6	8	28
15	7	9	31
16	8	10	34
17	9	11	37
18	10	12	40
19	11	13	43

19

PERENNIAL MATH

TEST # 3

March 2019 Rookie Level Solutions

4) In this cryptarithm, each letter stands for a different digit. What is the value of CRATE? First solve for the 3 E's. Since their sum ends in 8, E must be 6 and a 1 carries over to the tens digit. 3T's plus 1 ends in 2, so T must be 7 and a 2 carries over to the hundreds digit. 3 A's plus 2 ends in 4, so A must be 4 and a 1 carries over to the thousands digit. 3 R's plus 1 ends in 7, so R must be 2. There is no carry over, so C = 3. Therefore CRATE = 32476

```
   RATE
   RATE
+ CRATE
  37428
```

32476

5) Tom was bored in class so he started writing out the counting numbers beginning at 1 and then 2 and so on. What was the 185th digit that he wrote? So the pattern is to write out the counting numbers in an uninterrupted sequence. From 1 to 9 there are 9 digits, so that leaves us with 185 - 9 = 176 digits. This will leave us with 176/2 = 88 two-digit counting numbers remaining in the sequence. The 88th 2-digit number starting at 10 is 97. So the 185th digit is 7.

7

6) Gigi has a bag of 140 shells. The bag contains only limpets, cockles, and conchs. There are three times as many cockles as limpets. There are six times as many conchs as limpets. To find out how many conchs are in the bag make a chart. Start with 1 Limpet and increase it until there are 140 total shells. There were 84 conchs.

Limpets (L)	Cockles (3L)	Conchs (6L)	Total (140)
1	3	6	10
2	6	12	20
3	9	18	30
4	12	24	40
5	15	30	50
6	18	36	60
7	21	42	70
8	24	48	80
9	27	54	90
10	30	60	100
11	33	66	110
12	36	72	120
13	39	78	130
14	42	84	140

84

Name________________________________

PERENNIAL MATH

TEST #4

April 2019 Rookie Level

1) 144, 132, 122, 114, ___, ___, ___ What would be the sum of the three missing numbers that complete this pattern?

2) There were 2 algae put in the petri dish on Monday and by Tuesday morning there were four algae in the petri dish. If the number of algae double every day, how many algae will be in the petri dish by Sunday morning?

3) Tom has to do stretching exercises three times a day for fifteen minutes each time. He will have to do this for a week along with going to therapy for one hour exercises Monday through Friday. How many total *minutes* will he have exercised at the end of the week?

4) How many odd numbers between 400 and 500 are divisible by 7?

5) Four friends (Jack, John, Jill and Julie) line up. There are 2 boys and 2 girls. How many ways can they line up if the two girls cannot be beside each other?

6) Using the diagram below, what would the value of two circles plus one star?

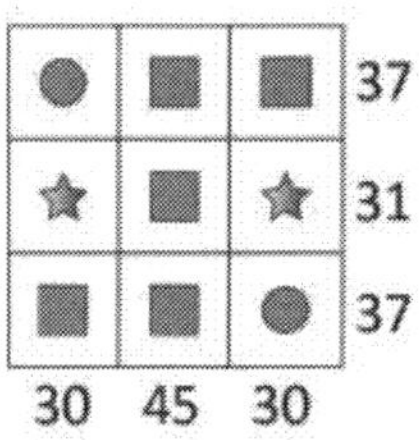

PERENNIAL MATH

TEST #4

April 2019 Rookie Level Answers

1) 144, 132, 122, 114, ___, ___, ___ What would be the sum of the three missing numbers that complete this pattern?	314
2) There were 2 algae put in the petri dish on Monday and by Tuesday morning there were four algae in the petri dish. If the number of algae double every day, how many algae will be in the petri dish by Sunday morning?	128
3) Tom has to do stretching exercises three times a day for fifteen minutes each time. He will have to do this for a week along with going to therapy for one hour exercises Monday through Friday. How many total *minutes* will he have exercised at the end of the week?	615
4) How many odd numbers between 400 and 500 are divisible by 7?	7
5) Four friends (Jack, John, Jill and Julie) line up. There are 2 boys and 2 girls. How many ways can they line up if the two girls cannot be beside each other?	12
6) Using the diagram below, what would the value of two circles plus one star? 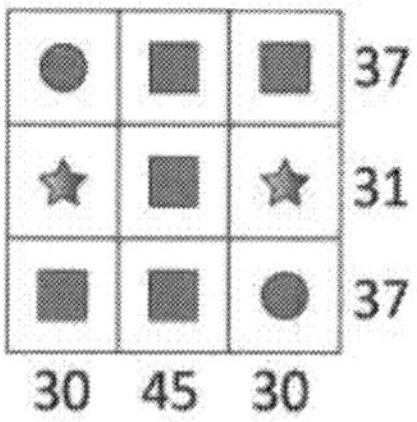	22

PERENNIAL MATH

TEST # 4

April 2019 Rookie Level Solutions

1) 144, 132, 122, 114, ___, ___, ___ . The pattern is a decrease in each number by -12, -10, -8, -6, -4, -2. The three missing numbers would be (114 - 6), (108 - 4), (104 - 2). The sum of the three missing numbers would be: 108 + 104 + 102 = 314.

314

2) There were 2 algae put in the petri dish on Monday and by Tuesday morning there were four algae in the petri dish. If the number of algae double every day, how many algae will be in the petri dish by Sunday morning? Use a chart to solve this problem. There will be 128 algae on Sunday morning.

Day	Start	End
Monday	2	4
Tuesday	4	8
Wednesday	8	16
Thursday	16	32
Friday	32	64
Saturday	64	128
Sunday	128	256

128

3) Tom has to do stretching exercises three times a day for fifteen minutes. He will have to do this for six weeks along with going to therapy for one hour Monday through Friday. To find the total *minutes* will he have exercised at the end of the six weeks, first calculate how many minutes he stretches at home per day. Three times fifteen equals 45 minutes per day. He does this for 1 week and there are 7 days in a week so he will stretch 45 minutes for 7 days. That is 315 minutes of stretching. Now we must add the therapy hours for the 5 days which is 60 minutes x 5 days = 300 minutes. His total minutes of stretching exercises for the week are 315 + 300 minutes = 615 minutes.

615

PERENNIAL MATH

TEST #4

April 2019 Rookie Level Solutions

4) How many odd numbers between 400 and 500 are divisible by 7? The first obvious multiple of 7 is 420. If you subtract 7 to get the preceding numbers down to 400, you get <u>413</u> and 406. Now just add 7 until you find the remaining multiples of 7: <u>427</u>, 434, <u>441</u>, 448, <u>455</u>, 462, <u>469</u>, 476, <u>483</u>, 490, <u>497</u>. There are 7 <u>odd</u> multiples of seven between 400 and 500.

7

5) Four friends line up. There are 2 boys and 2 girls. How many ways can they line up if the two girls cannot be beside each other? Act this out. You can also use letters to represent each person and make a list.

12

6) Use algebra to find the value of two circles plus one star. Let R = square; C = circle; and S = star.
First solve for the square (middle column). 3R = 45. 45/3 = 15 so R = 15.
Next solve for the value of the circle (top row). 2R + C = 37. So 2(15) + 7 = 37, and C = 7.
Next solve for the value of the star (middle row). 2S + R = 31. 2S + 15 =31, so 2S = 16, S = 8.
Now substitute the values for C and S in the equation 2C + S to solve the problem. 2(7) + (8) = 14 + 8 = 22.

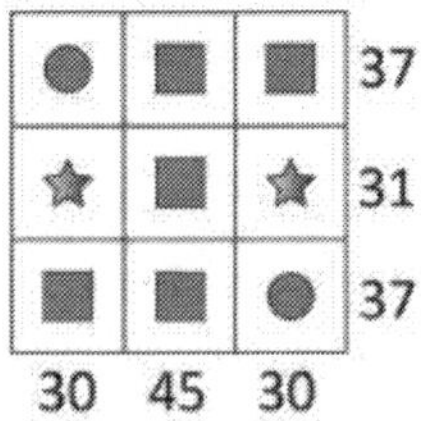

22

Name________________________________

PERENNIAL MATH

TEST #1

November 2019 Rookie Level

1) 25 + 87 = 267 – ____

2) A chess club starts the first meeting with 4 members. If each member brings 2 new members to the next meeting and then each member keeps bringing in 2 new members to each meeting, how many total members will there be at the fourth meeting?

3) There are 2 orange, 4 blue, 2 green, 3 yellow, and 12 red marbles. How many more red marbles will you need to add so that there are three times as many red marbles as all the other colors added together?

4) How many 3-digit numbers can be made that meet all these requirements:

- The digit in the ones place is greater than 7
- The digit in the tens place is less than 3
- The digit in the hundreds place is even
- The digits add up to an odd number

5) A rectangle is formed by placing two identical squares side by side. The perimeter of the rectangle is 36 inches. What is the area of the rectangle?

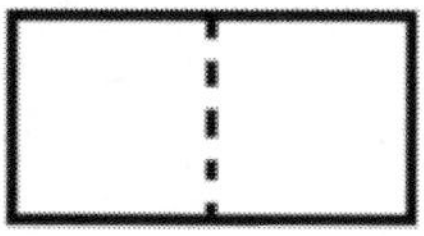

6) At the football game, the first group to arrive consisted of 4 people. Each group that arrived had twice the amount of people than the group before them. How many people in total, arrived in the first 6 groups?

PERENNIAL MATH TEST #1

November 2019 Rookie Level Answers

1) 25 + 87 = 267 – ____	**155**
2) A chess club starts the first meeting with 4 members. If each member brings 2 new members to the next meeting and then each member keeps bringing in 2 new members to each meeting, how many total members will there be at the fourth meeting?	**108**
3) There are 2 orange, 4 blue, 2 green, 3 yellow, and 12 red marbles. How many more red marbles will you need to add so that there are three times as many red marbles as all the other colors added together?	**21**
4) How many 3-digit numbers can be made that meet all these requirements: • The digit in the ones place is greater than 7 • The digit in the tens place is less than 3 • The digit in the hundreds place is even • The digits add up to an odd number	**12**
5) A rectangle is formed by placing two identical squares side by side. The perimeter of the rectangle is 36 inches. What is the area of the rectangle? 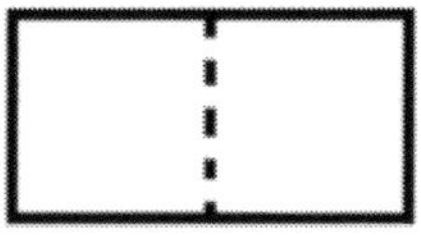	**72**
6) At the football game, the first group to arrive consisted of 4 people. Each group that arrived had twice the amount of people than the group before them. How many people in total, arrived in the first 6 groups?	**252**

PERENNIAL MATH TEST # 1

November 2019 Rookie Level Solutions

1) 25 + 87 = 267 – ____
= 267 – (25 + 87)
= 257 – 112 = 155

155

2) A chess club starts the first meeting with 4 members. If each member brings 2 new members to the next meeting and then each member keeps bringing in 2 new members to each meeting, how many total members will there be at the fourth meeting?
Make a chart. By the end of the fourth meeting there are 108 members.

Meeting	Members	+ 2 New Participants Each	Total Attendance
1	4	0	4
2	4	8	12
3	12	24	36
4	36	72	108

108

3) There are 2 orange, 4 blue, 2 green, 3 yellow, and 12 red marbles. How many more red marbles will you need to add so that there are three times as many red marbles as all the other colors added together? The total number of marbles that are not red are 2 + 4 + 2 + 3 = 11. Three times this amount is 33. 33 – 12 = 21. You will need to add 21 red marbles to have three times as many marble.

21

PERENNIAL MATH

TEST # 1

November 2019 Rookie Level Solutions

4) How many 3-digit numbers can be made that meet all these requirements:

- The digit in the ones place is greater than 7 {8,9}
- The digit in the tens place is less than 3 {0, 1, 2}
- The digit in the hundreds place is even {2, 4, 6, 8}
- The digits add up to an odd number

Possible numbers include: 209, 218, 229, 409, 418, 429, 609, 618, 629, 809, 818, 829.

12

5) A rectangle is formed by placing two identical squares side by side. The perimeter of the rectangle is 36 inches. What is the area of the rectangle? First solve for the dimensions of the rectangle. The length must be twice the width. The width is 6 and the length is 12 because 2 (12 + 6) = 36. The area is 6 x 12 = 72 inches^2.

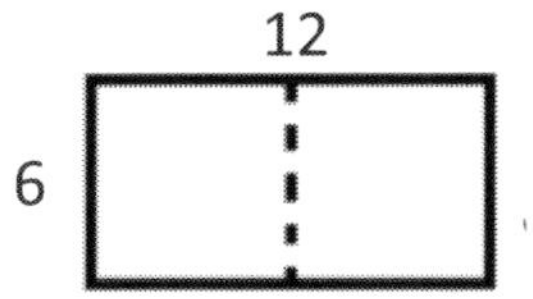

72

6) At the football game, the first group to arrive consisted of 4 people. Each group that arrived had twice the amount of people than the group before them. How many people in total, arrived in the first 6 groups? Make a chart. There were 252 people that arrived in the first 6 groups.

Group	Members
1	4
2	8
3	16
4	32
5	64
6	128
Total	252

252

Name________________________________

PERENNIAL MATH

TEST #2

December 2019 Rookie Level

1) 999 – 111 = 389 + _____	
2) Mark had $82 to spend on 7 tickets. After buying the tickets, he had $19 dollars left to spend at the event. How much did each ticket cost?	
3) According to the pattern below, how many consecutive odd integers are required to make a sum of 169? 1 + 3 = 4 1 + 3 + 5 = 9 1 + 3 + 5 + 7 = 16 1 + 3 + 5 + 7 + 9 = 25	
4) Peter, Jack, and Mary brought candy canes to the party. Peter brought three times as many as Jack. Mary brought twice as many as Peter. If Mary brought 18 candy canes, how many did Jack bring?	
5) Vince and Demetri have a total of 39 marbles. If Demetri has 7 more marbles than Vince, how many marbles does Vince?	
6) Valerie had $35 and withdrew additional money from her bank account. She bought a pair of shorts at $34, two shirts at $16 each and 2 pairs of shoes at $24 each. After shopping, she had $32 left. How much money did Valerie withdraw from the bank?	

PERENNIAL MATH

TEST #2

December 2019 Rookie Level Answers

1) 999 – 111 = 389 + _____	**499**
2) Mark had $82 to spend on 7 tickets. After buying the tickets, he had $19 dollars left to spend at the event. How much did each ticket cost?	**9**
3) According to the pattern below, how many consecutive odd integers are required to make a sum of 169? 1 + 3 = 4 1 + 3 + 5 = 9 1 + 3 + 5 + 7 = 16 1 + 3 + 5 + 7 + 9 = 25	**13**
4) Peter, Jack, and Mary brought candy canes to the party. Peter brought three times as many as Jack. Mary brought twice as many as Peter. If Mary brought 18 candy canes, how many did Jack bring?	**3**
5) Vince and Demetri have a total of 39 marbles. If Demetri has 7 more marbles than Vince, how many marbles does Vince?	**16**
6) Valerie had $35 and withdrew additional money from her bank account. She bought a pair of shorts at $34, two shirts at $16 each and 2 pairs of shoes at $24 each. After shopping, she had $32 left. How much money did Valerie withdraw from the bank?	**111**

PERENNIAL MATH — TEST #2

December 2019 Rookie Level Solutions

1) 999 – 111 = 389 + _____

888 = 389 + _____ {subtract 339 from both sides}

888 – 339 = 549

549

2) Mark had $82 to spend on 7 tickets. After buying the tickets, he had $19 dollars left to spend at the event. How much did each ticket cost?

First determine the total amount he spent. $82 - $19 = $63

Divide the total amount spent by the number of tickets. $63 / 7 = $9

Each ticket costs $9.

9

3) According to the pattern below, how many consecutive odd integers are required to make a sum of 169? Add up the odd numbers until you reach 169. There are 13 odd numbers that add up to 169.

1 + 3 = 4
1 + 3 + 5 = 9
1 + 3 + 5 + 7 = 16
1 + 3 + 5 + 7 + 9 = 25
25 + 11 = 36
36 + 13 = 49
49 + 15 = 64
64 + 17 = 81
81 + 19 = 100
100 + 21 = 121
121 + 23 = 144
144 + 25 = 169 (This is the 13th odd number).

13

PERENNIAL MATH

TEST # 2

December 2019 Rookie Level Solutions

4) Peter, Jack, and Mary brought candy canes to the party. Peter brought three times as many as Jack. Mary brought twice as many as Peter. If Mary brought 18 candy canes, how many did Jack bring? Since Mary brought twice as many as Peter and 18/2 = 9, then Peter brought 9 candy canes. If Peter brought three times as many as Jack, then 9/3 = 3. Therefore, Jack brought 3 candy canes.

3

5) Vince and Demetri have a total of 39 marbles. If Demetri has 7 more marbles than Vince, how many marbles does Vince? Subtract 7 from the total of 39 and divide the marbles in half. 39 – 7 = 32. 32/2 = 16. If Demetri has 7 more marbles, then he has 16 + 7 = 23 marbles. Vince has 16 marbles.

16

6) Valerie had $35 and withdrew additional money from her bank account. She bought a pair of shorts at $34, two shirts at $16 each and 2 pairs of shoes at $24 each. After shopping, she had $32 left. How much money did Valerie withdraw from the bank? Add up the items she bought: $34 + 2($16) + 2($24) = $34 + $32 + $48 = $114. Now subtract the money left from the money she started with $35 - $32 = $3. Add this amount to total spent, $114 - $3 = $111. She withdrew $111.

111

Name________________________________

PERENNIAL MATH

TEST #3

January 2020 Rookie Level

1) What is the sum of the first 5 even numbers greater than 56?

2) A plumber charges $225 for materials and $35 per hour for labor. The total cost of for the bill is $330. How many hours did it take the plumber to do the job?

3) A large box contains 18 small boxes and each small box contains 25 chocolate bars. How many chocolate bars are in the large box?

4) In this cryptarithm, different letters represent different digits. What is the three-digit number ABC?

$$
\begin{array}{r}
A\,B\,C \\
B\,C \\
+\quad C \\
\hline
A\,A\,A
\end{array}
$$

5) The teacher wants the class to collect 300 bottles of water for a community project. You brought in three bags that have six bottles in each bag. Your friend brings in six bags with six bottles in each bag. How many does the class still need to collect?

6) A zookeeper went to deliver some elephants and birds. He counted 28 legs. There were at least 3 of each type animal. What is the greatest number of elephants he could have counted?

PERENNIAL MATH

TEST #3

January 2020 Rookie Level Answers

1) What is the sum of the first 5 even numbers greater than 56?	310
2) A plumber charges $225 for materials and $35 per hour for labor. The total cost of for the bill is $330. How many hours did it take the plumber to do the job?	3
3) A large box contains 18 small boxes and each small box contains 25 chocolate bars. How many chocolate bars are in the large box?	450
4) In this cryptarithm, different letters represent different digits. What is the three-digit number ABC? $\begin{array}{r} ABC \\ BC \\ +\quad C \\ \hline AAA \end{array}$	632 or 418
5) The teacher wants the class to collect 300 bottles of water for a community project. You brought in three bags that have six bottles in each bag. Your friend brings in six bags with six bottles in each bag. How many does the class still need to collect?	246
6) A zookeeper went to deliver some elephants and birds. He counted 28 legs. There were at least 3 of each type animal. What is the greatest number of elephants he could have counted?	5

PERENNIAL MATH

TEST #3

January 2020 Rookie Level Solutions

1) What is the sum of the first 5 even numbers greater than 56?

58 + 60 + 62 + 64 + 66 = 310

310

2) A plumber charges $225 for materials and $35 per hour for labor. The total cost of for the bill is $330. How many hours did it take the plumber to do the job? Subtract the cost of the materials from the bill and then solve for how many hours the plumber worked. $330 - $225 = $105. $105/$35 = 3. The plumber worked for 3 hours.

3

3) A large box contains 18 small boxes and each small box contains 25 chocolate bars. How many chocolate bars are in the large box?
18 boxes x 25 chocolate bars = 450 chocolate bars in the large box.

450

PERENNIAL MATH

TEST # 3

January 2020 Rookie Level Solutions

4) In this cryptarithm, different letters represent different digits. What is the three-digit number ABC? Use trial and error. C cannot be 0 because A cannot be 0. C cannot be 1, but it can be 2. If C = 2, then A = 6 and B = 3. 632 + 32 + 2 = 666. ABC = 632. Another possible solution is 418.

$$\begin{array}{r} A\,B\,C \\ B\,C \\ +\quad C \\ \hline A\,A\,A \end{array}$$

632 or 418

5) The teacher wants the class to collect 300 bottles of water for a community project. You brought in three bags that have six bottles in each bag. Your friend brings in six bags with six bottles in each bag. How many does the class still need to collect? You brought in 3 x 6 = 18 bottles. Your friend brought in 6 x 6 = 36 bottles. Together you brought in 18 + 36 = 54 bottles. 300 – 54 = 246 bottles still needed.

246

6) A zookeeper went to deliver some elephants and birds. He counted 28 legs. There were at least 3 of each type animal. What is the greatest number of elephants he could have counted? Start with the initial amount of legs for 3 animals each and then add more elephants to it. 3 x 2 = 6 legs from birds. Now add the most number of elephants possible. 28 – 6 = 22 legs. That leaves 22 legs to divide up from the elephants. 22/4 = 5 with a remainder of 2. Therefore, there were 5 elephants and 4 birds.

5

Name________________________________

TEST #4

February 2020 Rookie Level

1) If  = 17 in the Mayan culture, what does  equal?

2) Add the adjacent numbers together and write their sum in the block above them. What is the sum of all fifteen numbers in the pyramid?

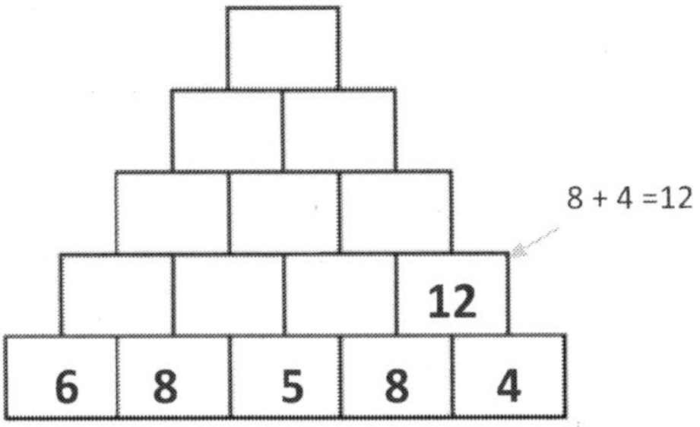

3) There are five teams in the tournament. If each team plays all of the other teams one time, how many games will occur?

4) A tour group of 64 members went to Chattanooga. Half of them went to the aquarium. They went through the aquarium in four equal groups. How many people were in each group?

5) Sheila bought a cookie for 45 cents. How many different ways could she have paid for it using dimes, nickels and pennies from her piggy bank? She must use all three coin types in her purchase.

6) Eric, Craig and Trevor are brothers. Craig is 5 years older than Eric. Trevor is 6 years older than Craig. The sum of their three ages is 31 years. How old is Eric?

PERENNIAL MATH

TEST # 4

February 2020 Rookie Level Answers

1) If  = 17 in the Mayan culture, what does  equal?

13, 18, or 23

2) Add the adjacent numbers together and write their sum in the block above them. What is the sum of all fifteen numbers in the pyramid?

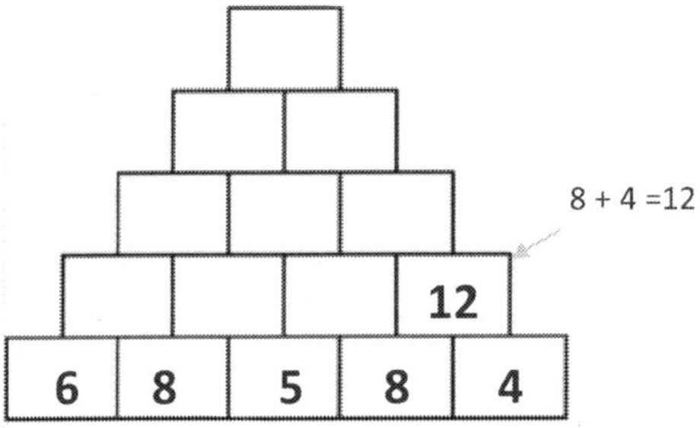

369

3) There are five teams in the tournament. If each team plays all of the other teams one time, how many games will occur?

10

4) A tour group of 64 members went to Chattanooga. Half of them went to the aquarium. They went through the aquarium in four equal groups. How many people were in each group?

8

5) Sheila bought a cookie for 45 cents. How many different ways could she have paid for it using dimes, nickels and pennies from her piggy bank? She must use all three coin types in her purchase.

12

6) Eric, Craig and Trevor are brothers. Craig is 5 years older than Eric. Trevor is 6 years older than Craig. The sum of their three ages is 31 years. How old is Eric?

5

PERENNIAL MATH

TEST #4

February 2020 Rookie Level Solutions

1) If  = 17 in the Mayan culture, what does  equal? One dash = 5 and one dot = 1. Therefore, 3 dots plus 2 dashes equals (3 x 1) + (2 x 5) = 13. Other alternate answers are that one dot = 4 and one dash = 3 and an answer of 18; and one dot = 7 and one dash = 1 and an answer of 23.

13, 18, or 23

2) Add the adjacent numbers together and write their sum in the block above them. To find the sum of all fifteen numbers in the pyramid, first solve for the numbers in the pyramid. 104 + 53 + 51 + 27 + 26 + 25 + 14 + 13 + 13 + 12 + 6 + 8 + 5 + 8 + 4 = 369.

		104		
	53		51	
	27	26	25	
14	13		13	12
6	8	5	8	4

369

3) There are five teams in the tournament. If each team plays all of the other teams one time, how many games will occur? Make a chart to show the team match ups. There will be a total of 4 + 3 + 2 + 1 = 10 games played.

Team 1	Team 2	Team 3	Team 4
Team 2	Team 3	Team 4	Team 5
Team 3	Team 4	Team 5	1 game
Team 4	Team 5	2 games	
Team 5	3 games		
4 games			

10

PERENNIAL MATH

TEST #4

February 2020 Rookie Level Solutions

4) A tour group of 64 members went to Chattanooga. Half of them went to the aquarium. They went through the aquarium in four equal groups. How many people were in each group? First divide 64 in half. 64/2 = 32. Of the 32 members going to the Aquarium, they were put in four equal groups. 32/4 = 8 members to each group.

8

5) Sheila bought a cookie for 45 cents. How many different ways could she have paid for it using dimes, nickels and pennies from her piggy bank? She must use all three coin types in her purchase. Make an organized chart. Start with the most amount of dimes and work down. There are 12 ways.

D	N	P	Total
3	2	5	45
3	1	10	45
2	4	5	45
2	3	10	45
2	2	15	45
2	1	20	45
1	6	5	45
1	5	10	45
1	4	15	45
1	3	20	45
1	2	25	45
1	1	30	45

12

6) Eric, Craig and Trevor are brothers. Craig is 5 years older than Eric. Trevor is 6 years older than Craig. The sum of their three ages is 31 years. How old is Eric? Make a chart to find Eric's age. Start with Eric as 1 year old and increase his age by 1 until the sum of their ages is 31. Eric is 5 years old.

Eric	Craig	Trevor	Total
1	6	12	19
2	7	13	22
3	8	14	25
4	9	15	28
5	10	16	31

5